NUMBER TRACING BOOK FOR PRESCHOOLERS

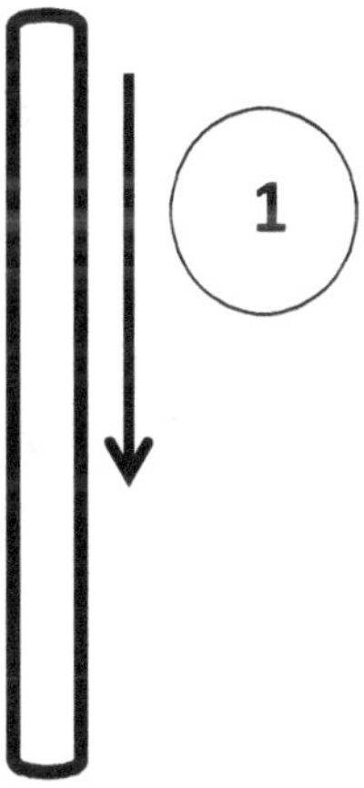
1

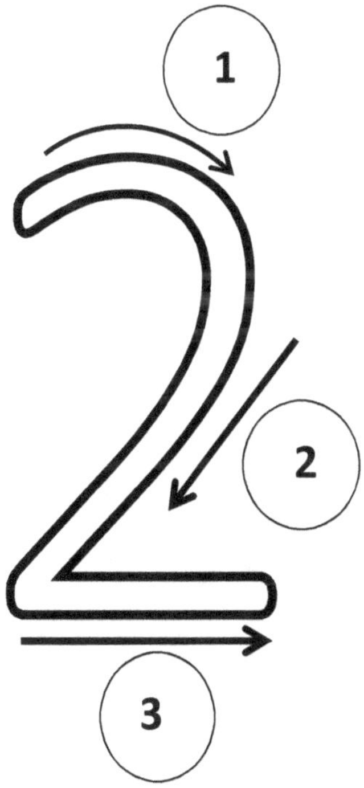
1
2
3

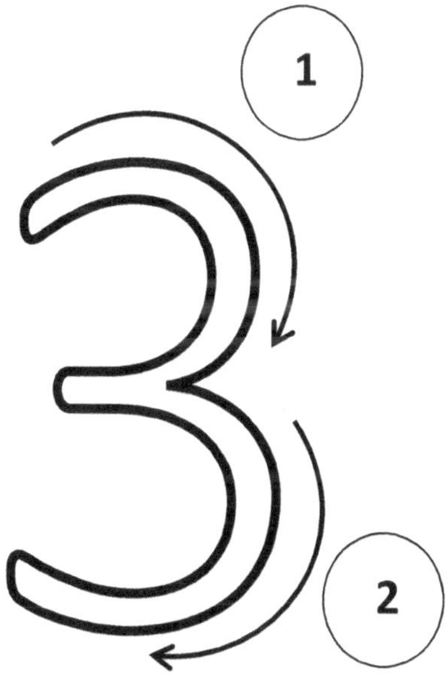
1
2

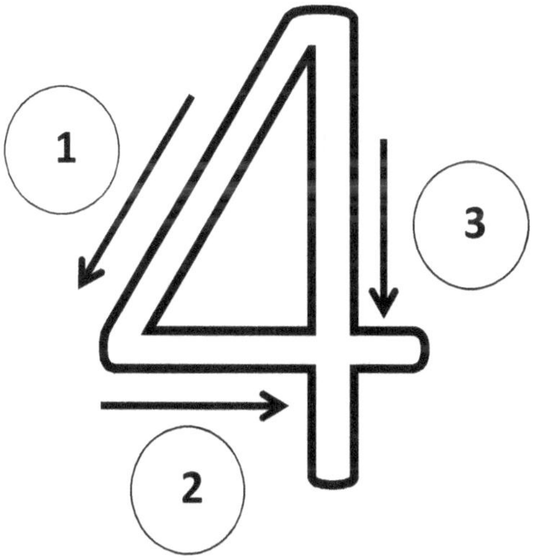
1
3
2

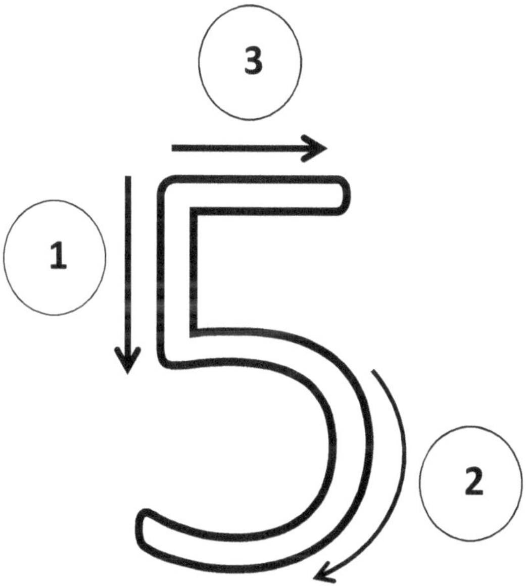
3
1
2

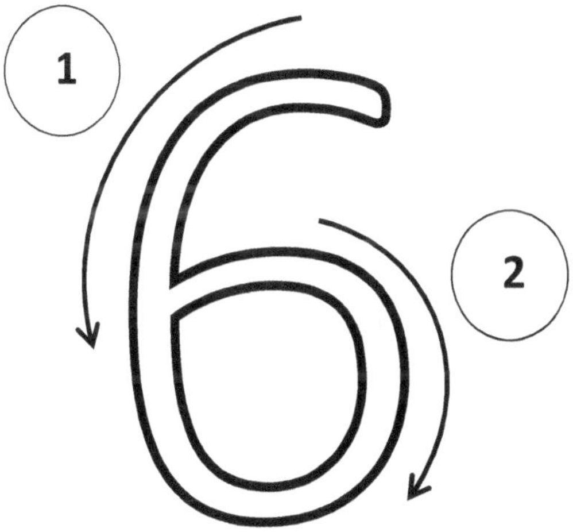
1
2

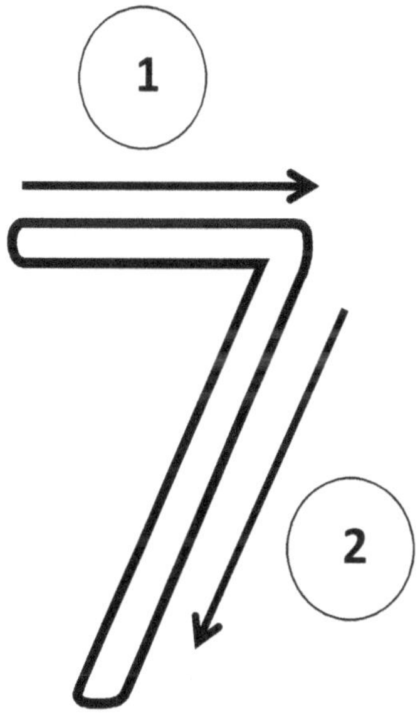
1
2

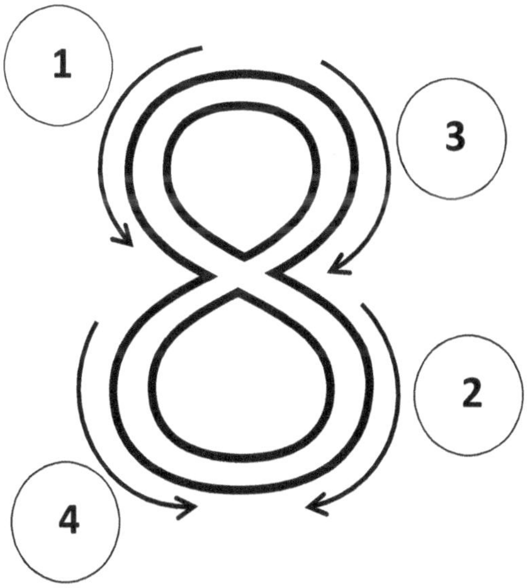
1
3
2
4

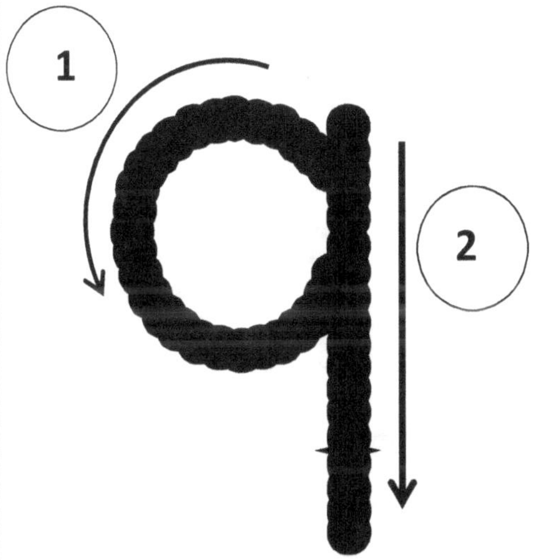
1
2

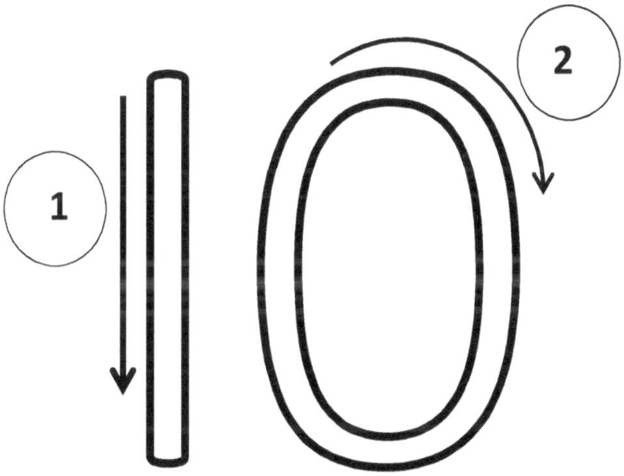
1
2

www.ingramcontent.com/pod-product-compliance
Lightning Source LLC
Chambersburg PA
CBHW081622220526
45468CB00010B/2987
* 9 7 8 1 7 2 1 1 2 7 0 1 6 *